STUDENT WORKBOOK TO ACCOMPANY
Small Animal Care
an ... ent

STUDENT WORKBOOK TO ACCOMPANY
Small Animal Care and Management

Third Edition

Dean M. Warren

Created by
Vicki Judah AS, CVT

DELMAR
CENGAGE Learning

Australia • Brazil • Japan • Korea • Mexico • Singapore • Spain • United Kingdom • United States

Title: Student Workbook to Accompany Small Animal Care and Management, Third Edition
Author(s): Dean M. Warren

Vice President, Career and Professional Editorial:
Dave Garza
Director of Learning Solutions: Matthew Kane
Managing Editor: Marah Bellegarde
Product Manager: Christina Gifford
Editorial Assistant: Scott Royael
Vice President, Career and Professional Marketing:
Jennifer McAvey
Marketing Director: Deborah Yarnell
Marketing Manager: Erin Brennan
Marketing Coordinator: Jonathan Sheean
Production Director: Carolyn Miller
Production Manager: Andrew Crouth
Content Project Manager: Anne Sherman
Art Director: David Arsenault

For product information and technology assistance, contact us at
Cengage Learning Customer & Sales Support, 1-800-354-9706

For permission to use material from this text or product,
submit all requests online at **www.cengage.com/permissions**
Further permissions questions can be e-mailed to
permissionrequest@cengage.com

Library of Congress Control Number: **2009921854**

ISBN-13: 978-1-4354-5337-1

ISBN-10: 1-4354-5337-9

Delmar
5 Maxwell Drive
Clifton Park, NY 12065-2919
USA

Cengage Learning is a leading provider of customized learning solutions with office locations around the globe, including Singapore, the United Kingdom, Australia, Mexico, Brazil, and Japan. Locate your local office at: **international.cengage.com/region**

Cengage Learning products are represented in Canada by Nelson Education, Ltd.

To learn more about Delmar, visit **www.cengage.com/delmar**

Purchase any of our products at your local college store or at our preferred online store **www.ichapters.com**

NOTICE TO THE READER
Publisher does not warrant or guarantee any of the products described herein or perform any independent analysis in connection with any of the product information contained herein. Publisher does not assume, and expressly disclaims, any obligation to obtain and include information other than that provided to it by the manufacturer. The reader is expressly warned to consider and adopt all safety precautions that might be indicated by the activities described herein and to avoid all potential hazards. By following the instructions contained herein, the reader willingly assumes all risks in connection with such instructions. The publisher makes no representations or warranties of any kind, including but not limited to, the warranties of fitness for particular purpose or merchantability, nor are any such representations implied with respect to the material set forth herein, and the publisher takes no responsibility with respect to such material. The publisher shall not be liable for any special, consequential, or exemplary damages resulting, in whole or part, from the readers' use of, or reliance upon, this material.

Printed in the United States of America
4 5 6 7 8   16 15 14 13 12

FD249

CONTENTS

SECTION 1

CHAPTER 1

Introduction to Small Animal Care

1. To help you become more familiar with some of the terms in this chapter, fill in the blanks in the following sentences.

    (a) Dinosaurs and marine reptiles were most abundant during the _____ period.

    (b) _____ _____ is credited with developing the system for the naming of species.

    (c) The system for the classification of species is _____.

    (d) The naming of species with this system is _____.

2. Rearrange the following words and put in them in the correct taxonomic sequence:

    phylum, species, order, kingdom, family, class, genus

    _____ _____ _____ _____ _____

    _____ _____

3. List four characteristics of mammals.

    (a)

    (b)

    (c)

    (d)

4. With rare exception, all mammals give birth to live young. However, there are two species of mammals that lay eggs. You will need to research your answers.

    (a) _____

    (b) _____

Name _____

5. Complete the Crossword Puzzle relating to the information in Chapter 1.

Crossword Small Animal Care

<u>Across</u>

2. The taxonomic classification for birds
4. Animals without backbones
5. The scientific naming of species
8. The embryonic stage of the vertebra
9. An atomic form
10. Layered structures created by algae and bacteria
12. Invertebrates with jointed legs
13. Single-celled animals including algae

<u>Down</u>

1. A scientist who deals with prehistoric life
3. Mammalian organ that nourishes a growing embryo
6. The oldest vertebrate group
7. Animals with a spinal cord
11. Bacterial organisms

CHAPTER 2
Safety

1. Many restaurants forbid animals from entering the premises. The exception to this is service animals. A service animal is one that provides assistance to the owner. What are some of the public health concerns of allowing animals in restaurants? Enter as many reasons as you can think of, adding more lines as necessary.

   (a) _____

   (b) _____

   (c) _____

   (d) _____

2. Refer to Figure 2-19 in your textbook. All species of ticks have common features. List three things you observe from the illustration.

   (a) _____

   (b) _____

   (c) _____

3. Complete the following sentences:

   (a) An _____ _____ is necessary for tapeworms to complete their life cycles.

   (b) Information regarding all aspects of a chemical can be found on the _____ provided with the product.

   (c) The method of controlling an animal for treatment or a procedure is referred to as _____.

   (d) The natural, immune host for many diseases is referred to as the _____.

   (e) The device used to prevent an animal from chewing or licking at a wound or incision site is called an _____ _____.

Name _____

4. Complete the Crossword Puzzle relating to the information in Chapter 2.

Crossword Safety

Across

5. Prefix for a disease caused by a tick

7. A wound, a cut or tear in the skin

9. Within the muscle

11. A device that provides protection from inhaling toxic substances

12. The only U.S. state that has never reported a case of rabies

13. An organism that lives on or within another organism

14. A zoonotic disease that can be transmitted from handling contaminated cat litter

Down

1. Device used to prevent an animal from biting

2. Diseases transmitted from animals to humans

3. Chemical that kills insects

4. The primary victims of animal bites

6. Most common intermediate host for *Echinococcus multiloculoaris*

8. The term used for controlling an animal

10. A skin disease caused by a fungus

Name _____

5. Complete the Word Scramble relating to the information in Chapter 2.

Word Scramble Safety

1. EFALR　　　　　— — — — —
2. RAVLA　　　　　— — — — —
3. AOSAMNLLEL　　— — — — — — — — — —
4. SEIONLVU　　　 — — — — — — — —
5. UCSNOCECOHCI　— — — — — — — — — — — —
6. SRAADIC　　　　— — — — — — —
7. MYHPN　　　　 — — — — —
8. AITCKISTER　　 — — — — — — — — — —
9. CETSSNENAU　　— — — — — — — — — —

Definitions for Word Scramble:

1. _____
2. _____
3. _____
4. _____
5. _____
6. _____
7. _____
8. _____
9. _____

CHAPTER 3

Small Animals as Pets

1. One of the suggested activities in your text is to become involved with a group that takes animals into nursing homes and hospitals. It is required that all animals pass specific behavioral tests before being approved, or certified, to enter into these programs. **What would you need to do to qualify your own pet as a therapy animal?** Using any search engine on the computer, enter "therapy dogs" as your topic. This should bring several results as a starting point. Follow the links on the selected Web sites to determine the process for qualification. You will need to consider not only behavioral concerns, but also research the animal's health requirements. Write your findings in the space provided. Use extra paper if required and attach the paper to the workbook.

   Behavioral: _____

   _____

   _____

   Health: _____

   _____

   _____

2. Visit your local animal shelter and view the animals available for adoption and answer the following:

   (a) What is the percentage of dogs to cats? _____

   (b) What breed of dog is most represented? _____

   (c) Does the shelter house any other small pets for adoption?

   _____

   (d) Determine the number one reason pets become available for adoption.

   _____

   (e) Request an application to adopt an animal. Carefully read the questions on the form, answering them as best you can. Attach this form to your workbook.

(f) Interview the staff and ask if there are volunteer positions available. In the space below, write their responses.

_____

_____

_____

_____

_____

3. What is the difference between a service animal and a therapy animal?

_____

4. Research your favorite breed of dog and determine what health concerns are associated with the breed. For example, your text cites **entropion** and the skin conditions prevalent in Shar-Pei dogs.

(a) Breed: _____

(b) Medical concerns: _____

_____

_____

_____

5. To help build your vocabulary, create a list and define the terms you encounter regarding medical conditions and terminology. Add additional sheets throughout the course and a build a reference source for yourself. Three common conditions have been listed below to help you get started.

Entropion _____

Cherry Eye _____

Hip Dysplasia _____

Name _____

6. Complete the Word Search for terms in Chapter 3.

Word Search Small Animals as Pets

```
W M C X D C F O S X M P J R W X B S Z L M U B L R X A
H Z W T H E R A P Y D O G S N N Q G R I E F A B D U D
O O Y I Q I I C U O M O H I N Y G C G P R K G S L D O
P L S G Y D N U P W H G U S B E C L Y A U I R H G A P
P B I A Q W U Q A D E T M Z C V A J R V P D Q C M P T
I F J J R Z O W H X A F A S T A T U S S Y M B O L S I
B I N O U D A V T Y D C N O F K D R L S U J R D G L O
T G M T P H B L Q E J I E C E C A I S X F A J O I M N
I U T W C H W N U U J C S U C D I Q I Z L X Y M I N F
T E U T H A N A S I A C O H W S A Z Q U H I U E O Q X
N E H N S F R E O F C F C N Y N E M C U S F B S T E L
S C U N A S E N I Q S I I Q P S E S H P O N A T E L S
E X W Q L Q S G O E L U E E N E A U M R O O A I V N W
R B T G L D P I N L V T T M E V X E T M Y K X C E G R
V L X Y E Y O X Z T E W Y R O J P U P E Y I V A Q R S
I Y K N R X N J L R R M G I G Q P W D V R L H T A K Q
C L H X G O S M Y W Y I D S K Y C G U E J L T E H A F
E S S D I A I P J P D R M U R J E R Q N B S S D E N Y
D I H A E L B G N E A A O E M C D X A T D H Y N R E K
O R H M S I I A P C U S P K P E C J J R H E E D E S V
G E B A A F L Q I R C F I A A D K X I O Z L M S D T T
S M S G K E I T C O M P A S S I O N C P H T Q P I H F
X U E P B S T P S D D R K A G R A G O I Q E W Q T E K
C M G C A T Y J J W I N L X G M L I F O T R S I A T W
O R Q Z K Y Q S G U V F L T X A K U M N A U M A R I N
M Y U W R L F S M O E T M Z A R U M B V C U V Q Y Z R
S X U E X E T H Y P E R T E N S I O N P Z R Q T T E Q
```

Entropion	Euthanasia	Allergies	Neuter
Spay	Sire	Dam	Hereditary
No Kill Shelter	Grief	Hypertension	Cardiovascular
Anesthetize	Tumor	Service Dogs	Therapy Dogs
Responsibility	Compassion	Pedigree	Status Symbol
Domesticated	Lifestyle	Humane Society	Adoption

CHAPTER 4

Animal Rights and Animal Welfare

Some of the exercises or projects in this chapter of your workbook require reflective writing and ask you to explore your own thoughts regarding the issues presented in Chapter 4. You may already have had some experiences to help you form an opinion, or perhaps you have not really thought about some of the issues presented.

1. Visit the local public library and select a book that discusses animal rights or animal welfare. If the selection is small, ask a librarian to help you find suitable books that can be transferred to your branch. After reading the book you will need to write a report to be turned in to your instructor. It should include a brief summary of the contents, the author's position (either animal welfare or animal rights), the author's argument, whether you agree or disagree, and your conclusion.

2. Visit a large chain drug store that sells multiple household items. Before you begin, ask to speak to the store manager and explain what you would like to do and why. Ask permission first. If permission is given, compile a list of five of the most popular items in each section. Write the name of the company, the product, and, from the label, determine if the product has been tested on animals. If it has *not* been, there will be a statement such as "Cruelty Free" or "No Animal Testing." If there is no statement, assume that the product has been tested on animals. With your list of products that have been animal tested, look for alternatives to the same type of products.

   (a) Are cruelty-free products readily available as an alternative?

   (b) Does the same manufacturer produce both products?

   (c) Compare the price and value of those items that have been tested on animals to those that have not been. From the data you gather, create a chart that reflects your findings. Headings should include Product and Manufacturer, Alternative Product and Manufacturer, and Cost Comparison.

   (d) Write up your findings and compare the information with that of other students.

3. The introduction material of many chapters in your textbook contains information regarding the contributions made by various species in the advancement of human medicine and health. List the species and the different ways research with these animals has benefited human medicine. Compile your list separately and add to this by further searching the Internet. Keywords to search include: animal contributions to human health. Add your paper to the workbook for further reference.

4. Most people have conflicts in determining a balance between being either *for* or *against* animal research, animal rights, or animal welfare. All would probably support animal welfare, but there is conflict between the two. It is not easy, it is not "black and white," and no answer is either right or wrong. On a separate sheet of paper, list your personal concerns. For example: It may be acceptable to you if animals are used in research to benefit advancements in human medicine, but it is not acceptable to test human products on animals. If you don't wish to share this information, when you complete the assignment, mark it off in the space provided.

   Completed _____ Date _____

Name _____

5. Complete the Word Scramble relating to the information in Chapter 4.

Word Scramble Animal Rights and Animal Welfare

1. PTAE   __ __ __ __
2. NLAAIM HSTIGR   __ __ __ __ __ __ __ __ __ __ __ __
3. LMNAAI AERWEFL   __ __ __ __ __ __ __ __ __ __ __ __
4. IROEORESCMTR   __ __ __ __ __ __ __ __ __ __ __ __
5. MNHUEIAZ   __ __ __ __ __ __ __ __
6. BHRGE   __ __ __ __ __
7. ISNVCTOIEVI   __ __ __ __ __ __ __ __ __ __
8. NLAMIA RAILEOBNTI NROFT   __ __ __ __ __ __ __ __ __ __ __ __ __ __ __ __ __ __ __ __
9. CLTNUEIAH   __ __ __ __ __ __ __ __ __
10. NI ITVOR   __ __ __ __ __ __ __
11. ERISGN   __ __ __ __ __ __
12. BRASIBT   __ __ __ __ __ __ __
13. EEPCMSISI   __ __ __ __ __ __ __ __ __
14. COBP   __ __ __ __
15. EVLA   __ __ __ __

CHAPTER 5

Careers in Small Animal Care

1. Imagine yourself employed in the ideal job, the one that you have always worked toward and have envisioned yourself doing.

    (a) Describe this job: _____

    _____

    _____

    _____

    (b) What education did your receive to qualify you for the position?

    _____

    _____

    _____

    _____

    (c) How did you find this position? _____

    _____

    _____

2. From the information provided in Chapter 5, make a list of the positions in animal care that most closely resemble the ideal career you have described above.

    (a) What are some of the things these careers have in common?

    _____

    _____

    _____

    _____

(b) From the list of careers that most match your own ideal job, write down all of the things that you could be doing now to help you obtain the work you desire in the future.

_____

_____

_____

_____

_____

(c) Give the career name of your ideal work and define it by a title.

_____

3. Visit the American Zoo and Aquaruim Web site (http://www.AZA.org). The home page has subheadings. Click the tab for Job Openings. Here you will find an extensive list of available positions throughout the country and occasionally abroad. On the far left is the title of the position.

Click on this and it will provide complete information regarding the job description, qualifications, benefits and rate of pay, and how to submit an application. Complete the questions and projects below.

(a) What is an externship? _____

(b) What is a stipend? _____

(c) Select 10 openings that are of interest and, using separate sheets of paper, list the job details. Compare these specific positions to your answers for questions 1 and 2 above.

(d) How many of these would you qualify for upon graduation from high school? _____ How many others require advanced education? _____

(e) In addition to a formal education, what other qualification are sought?

_____

_____

_____

(f) List the other qualifications and highlight those you could obtain while still in school (for example, SCUBA certification).

_____

_____

_____

_____

(g) List the job requirements that surprised you the most.

_____

_____

_____

5. For any employment, correct terminology is important. Define the suffixes below:

(a) *-ology* _____

(b) *-ist* _____

(c) *-ician* _____

Sample words: cosmet*ology*, geolog*ist*, pediatr*ician*.

Name _____

6. Unscramble and define the following terms.

Word Scramble Careers in Small Animals

1. CSGINEET   __ __ __ __ __ __ __ __
2. SRUHNYDBA   __ __ __ __ __ __ __ __ __
3. HNIMETRCNE   __ __ __ __ __ __ __ __ __ __
4. MTHPAYE   __ __ __ __ __ __ __
5. IONNLCG   __ __ __ __ __ __ __
6. LIAOGTPOTSH   __ __ __ __ __ __ __ __ __ __ __
7. SLIOITIMOCGRBO   __ __ __ __ __ __ __ __ __ __ __ __ __ __
8. YOILYHSPGO   __ __ __ __ __ __ __ __ __ __
9. ITNNPISRHE   __ __ __ __ __ __ __ __ __ __

Definitions for Word Scramble:

1. _____
2. _____
3. _____
4. _____
5. _____
6. _____
7. _____
8. _____
9. _____

CHAPTER 6

Nutrition and Digestive Systems

1. Review the vitamins in this chapter. Using poster board, create a chart that lists each vitamin, its importance to health, and the diseases or conditions caused by an insufficient dietary intake.

   Your first entry may look like this:

Vitamin	Importance	Disease or Condition
A	Vision, respiration, reproduction	Poor vision, respiratory problems, difficulties with reproduction

   When you have completed the vitamins, add the same information for minerals.

   You may wish to enhance the information on your chart with research on the Internet. This will give you a visual summary of the importance of vitamins and minerals.

2. The prefix *hyper-* indicates a condition of excess. For example, ingestion of too much of a vitamin results in a condition of hypervitaminosis.

   What are the two most common vitamins given in excess?

   (a) _____

   (b) _____

3. As illustrated in your text, there are differences in animal digestive systems. Without referring to your book, label the indicated areas for the following digestive systems:

   The ruminant digestive system:

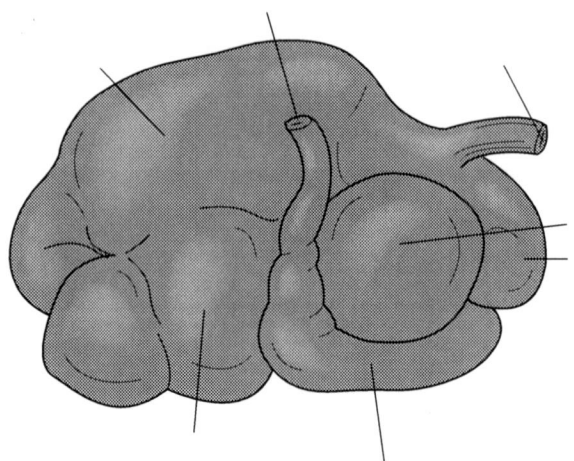

A monogastric digestive system:

The digestive system of a bird:

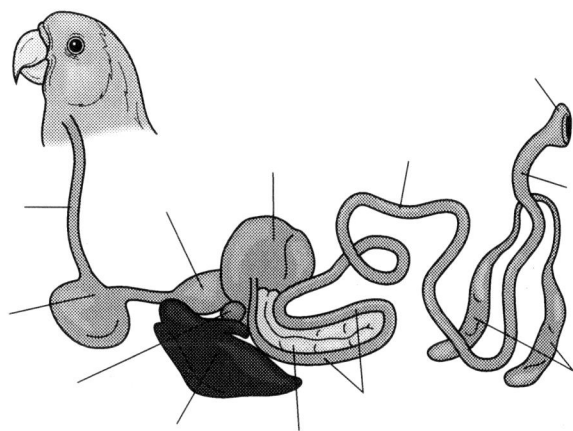

4. Horses and rabbits are quite different species. What organ is present in both species that enables them to convert roughage into digestible carbohydrates?

_____

5. What does this common organ contain to facilitate this process?

_____

Name _____

6. Complete the Crossword Puzzle relating to the information in Chapter 6.

Crossword Nutrition and Digestion

Across

3. Conversion of food so it can be digested
5. Foods that support life
7. Important for bone health
8. Disease associated with an iron deficiency
10. A component of carbohydrates
12. One of the two classifications of minerals
14. Animals with a four-chambered stomach
16. Another term for the gizzard
17. Present in every animal cell
18. Component in blood that carries oxygen

Down

1. Natural source of vitamin D
2. Necessary for the prevention of scurvy
3. The building blocks of proteins
4. Primary site for digestion
6. Contains the first digestive enzymes
8. The "true" stomach in cattle
9. How vitamins are classified
11. Required for the absorption of vitamins A, E, D, and K
13. Salt provides sodium and _____
15. Organ containing bacteria that break down roughage

SECTION 2

CHAPTER 7

Dogs

1. There are more than 120 breeds of dog recognized by various kennel clubs. When working with dogs in any capacity, it is important to know the different breeds. Although each dog may exhibit its own unique personality, there are certain characteristics of each breed. List five reasons why it is important to recognize the various dog breeds.

   (a) _____

   (b) _____

   (c) _____

   (d) _____

   (e) _____

2. You may be familiar with many of the breeds illustrated in your textbook, but it may seem a daunting task to know them all. A good method of study is to create flashcards for yourself or to share the task with teammates. Team up with six of your classmates, each of you creating the flashcards for one of the seven major groups of dogs. To obtain the breed photographs, access the Internet and use any search engine (keywords: dog breeds). Print the photos from the Web pages and glue them to one side of a 3 × 5 index card. On the other side of the card, write the name of the breed and its group. Create cards for only the dogs represented or mentioned in your textbook. Use the cards to quiz each other and keep them for future reference.

3. Dogs, Dogs, Dogs! Where in the world do they all come from? Obtain a large, poster-sized world map and display it on the bulletin board. From the information in your textbook, place one pin for each breed's country of origin. For example, the Bedlington Terrier originated in Great Britain. A pin should be placed in that country. Use different colored pins to designate the different groups. For example, use red pins for terriers, blue for hounds, and so forth.

   (a) Which country has developed the greatest number of breeds?

   _____

   (b) What are some of the reasons for such diversity in each country?

   _____

   _____

   _____

4. What are some of the benefits to spaying or neutering a dog?

    (a) Female: _____

    _____

    _____

    (b) Male: _____

    _____

    _____

Name _____

5. Unscramble the following list of dog diseases. When you have finished, list three signs of illness for each disease.

Word Scramble Dog Diseases

1. ONUVRSRCIOA   __ __ __ __ __ __ __ __ __ __ __

2. RITPOEISSPSOL   __ __ __ __ __ __ __ __ __ __ __ __ __

3. RPUVIOVASR   __ __ __ __ __ __ __ __ __ __

4. MRSDTEIPE   __ __ __ __ __ __ __ __ __

5. PATSIHETI   __ __ __ __ __ __ __ __ __

6. HNTEBICIOTSOCHRAR   __ __ __ __ __ __ __ __ __ __ __ __ __ __ __ __ __

7. NNAAPUAFZIERL   __ __ __ __ __ __ __ __ __ __ __ __ __

Unscrambled words and the signs of the disease.

1. _____

    (a) _____

    (b) _____

    (c) _____

2. _____

    (a) _____

    (b) _____

    (c) _____

3. _____

    (a) _____

    (b) _____

    (c) _____

4. _____

    (a) _____

    (b) _____

    (c) _____

5. _____

    (a) _____

    (b) _____

    (c) _____

6. _____

    (a) _____

    (b) _____

    (c) _____

7. _____

    (a) _____

    (b) _____

    (c) _____

Name _____

6. Complete the Crossword Puzzle relating to the information in Chapter 7.

Crossword Dog Breeds

Across

1. A terrier with the looks of a lamb
5. Group developed to hunt game
6. Descendant of Roman Drover Dogs
7. Once held sacred in China
11. Breed orginally known as the Badgerdog
13. The smallest of all breeds
14. A large sight hound from Russia
16. Often associated with fire trucks and carriages
17. Known for tenacity and going to ground

Down

1. Named after an American city
2. Breed made famous in a novel by Sir Walter Scott
3. A type of corgi
4. The smallest in size but not in affection
8. Breed with a blue-gray coat and golden eyes
9. Group developed to assist with livestock
10. Heavier in bone than any other breed
12. May need to be delivered by caesarean section
15. Some hunt by sight, some hunt by scent

CHAPTER 8

Cats

1. Access the Web pages for the two major cat breed organizations, CFA (Cat Fanciers Association) and TICA (The International Cat Association) and prepare flashcards for the breeds of cats represented in your textbook. These should be used in the same way as those you made for dog breeds.

2. Cats require a different diet from that fed to dogs. What are the two most important nutritional differences?

   (a) _____

   (b) _____

3. Some owners elect to have their cats declawed. Consult with veterinary staff and and discuss the pros and cons of this surgical procedure.

   _____

   _____

   _____

   _____

   Declawing of cats is controversial. From your notes above, discuss whether or not you would have your own cat declawed. Also consider your position on animals rights and animal welfare in your answer.

   _____

   _____

   _____

   _____

   _____

   _____

4. Research and write a paper on your favorite breed of cat. Describe what appeals to you about the breed, its unique characteristics, the history of the breed, and any medical or hereditary concerns associated with the breed. For further information, enter the breed of cat as your keyword search on the Internet. The Web sites of the CFA (Cat Fanciers Association) and TICA (The International Cat Association) will also provide specific information about your favorite breed. Present your report orally to the class.

Name _____

5. Complete the Word Scramble relating to the information in Chapter 8.

Word Scramble Cats

1. TCOHLAPIHM   _ _ _ _ _ _ _ _ _

2. NMAX   _ _ _ _

3. OMT   _ _ _

4. JCNOSASOB   _ _ _ _ _ _ _ _

5. ENUQE   _ _ _ _ _

6. AIETNA   _ _ _ _ _ _

7. FPI   _ _ _

8. ADNJCUIE   _ _ _ _ _ _ _

9. ARLEF   _ _ _ _ _

10. TKTRESAII   _ _ _ _ _ _ _ _

11. SOMTASLOIXSOP   _ _ _ _ _ _ _ _ _ _ _ _ _

12. ORLTSOMCU   _ _ _ _ _ _ _ _

13. OLCAIC   _ _ _ _ _ _

14. APLLPAEI   _ _ _ _ _ _ _ _

15. ICRIIVUALSC   _ _ _ _ _ _ _ _ _

16. RTUNAIE   _ _ _ _ _ _ _

17. CIITTNATIGN RNEBEAMM   _ _ _ _ _ _ _ _ _ _

    _ _ _ _ _ _ _

18. AEOCCHL   _ _ _ _ _ _ _

CHAPTER 9

Rabbits

1. In working with animals, you will often be confronted by the metric system. The metric system is most commonly used in drug dosages and in figuring dietary needs. Refer to Table 9-3 in your textbook. The table is based on the nutritional amount per kilogram of body weight. Determine the following:

   (a) Your rabbit weighs 3.5 pounds. What is its weigh in kilograms?

   _____

   (b) How many grams equal one kilogram?

   _____

   (c) If the rabbit weighs 54 ounces, what does it weigh in kilograms?

   _____

   (d) Round this number to the nearest decimal point.

   _____

2. Many new terms have been included in this chapter. To understand the exact meaning of these terms in general, determine the conditions the following words, prefixes, and suffixes indicate and add them to your growing list of terminology:

   (a) *entero-* _____
   (b) *toxemia* _____
   (c) *gastro-* _____
   (d) *hepatic* _____
   (e) *intra-* _____
   (f) *-osis* _____
   (g) *-cide* _____
   (h) *-itis* _____

3. Antibiotics are prescribed for bacterial infections. What defines a *broad-spectrum antibiotic*?

   _____

4. Many species sometimes exhibit the common behavior of teeth grinding. What is usually the cause of this behavior?

   _____

Name _____

5. Complete the Crossword Puzzle relating to the information in Chapter 9.

Crossword Rabbits

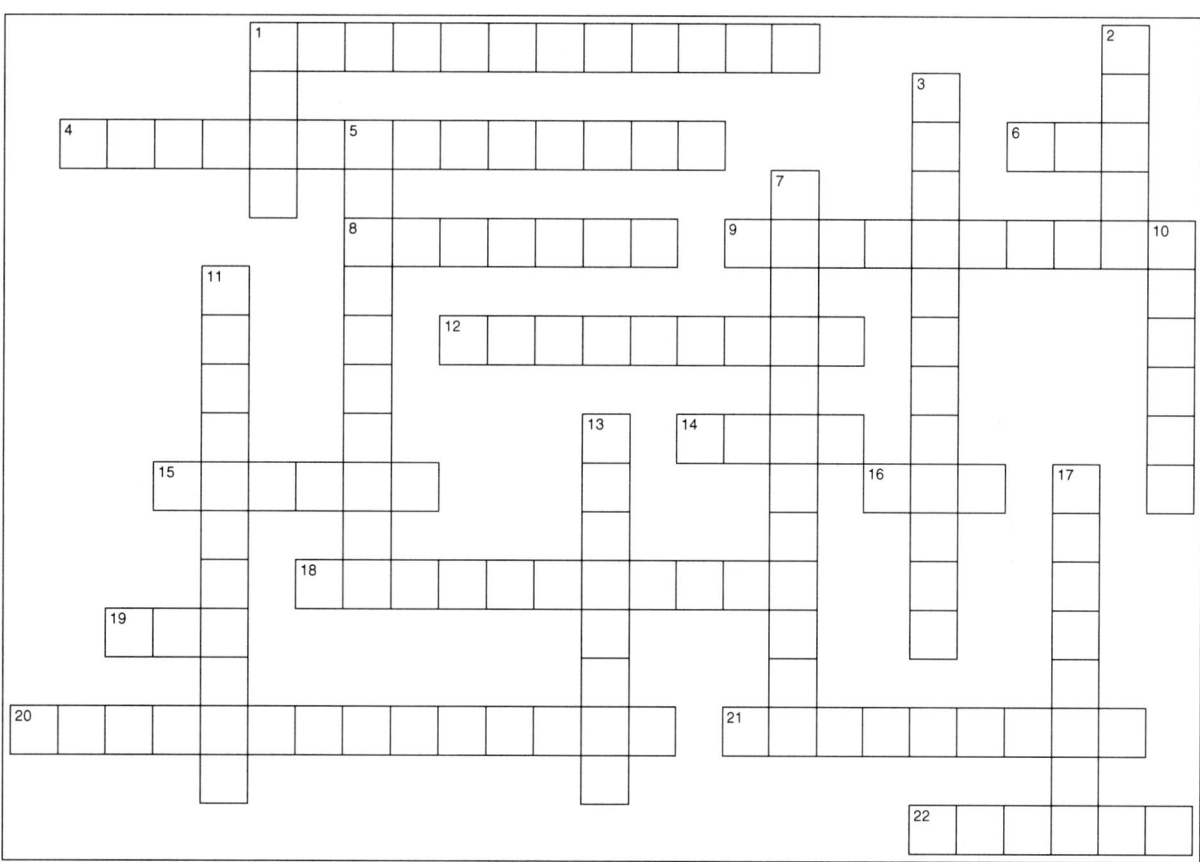

Across

1. Breed most commonly used in research
4. Another term for "Weepy Eye"
6. Term for a young rabbit
8. Common internal parasite of rabbits
9. The taxonomic classification
12. Method used to determine pregnancy
14. Term for a male rabbit
15. The roll of skin under the chin
16. Term for a young rabbit
18. Infectious fatal disease of rabbits caused by a virus
19. Term for a female rabbit
20. Condition of warts around the ears and mouth
21. Inflammation of the intestinal tract
22. How rabbit categories are determined

Down

1. The number of upper incisor teeth
2. Term used for rabbit housing
3. The largest breed of domestic rabbit
5. The consumption of "night feces"
7. The condition of having overgrown teeth
10. Breed with long, soft fur
11. Common bacterial infection of rabbits
13. Inflammation of the mammary glands
17. The term for a rabbit giving birth

CHAPTER 10

Hamsters

1. Assume that you have decided to keep a hamster as a pet. You have no supplies or appropriate housing. Visit a large retail pet store and compile a list of all the basic items you would need for the hamster. Write your items in the spaces below.

_____

_____

_____

_____

_____

   (a) What is the total cost of the items on your list? $_____

   (b) What is the cost of the hamster? $_____

   (c) Determine the real cost of having a hamster for a pet, assuming it is healthy and will reach its full life expectancy. $_____

2. Many pet stores offer a guarantee on the live animals they sell. Ask about the store's policy and enter the information below.

_____

_____

_____

3. The store provides for a replacement hamster if the one you purchase dies within a specified number of days. Within 2 days, your new pet develops very smelly diarrhea and is found dead in the cage.

    (a) What is the likely cause of death?

    _____

    (b) Would it be a good idea to obtain another hamster from the same colony?
    Why, or why not?

    _____

    _____

    _____

    _____

4. Access the Web site for the Centers for Disease Control and Prevention (http://cdc.org). Research the current information available on LCM (lymphocytic choriomeningitis). Make a special note of reported zoonotic cases and write a short paper on any updates. Attach this information to your workbook.

Name _____

5. Complete the Word Scramble relating to the information in Chapter 10.

Word Scramble Hamsters

1. SIEMNTGINI   __ __ __ __ __ __ __ __ __

2. RFADW   __ __ __ __ __

3. STEDORUXE   __ __ __ __ __ __ __ __ __

4. CRAATIEB   __ __ __ __ __ __ __ __

5. EHCKE OEPHSUC   __ __ __ __ __ __ __ __ __ __ __

6. WODO VGSSNIAH   __ __ __ __ __ __ __ __ __ __ __ __

7. ATWRE ETTBLO   __ __ __ __ __ __ __ __ __ __ __

8. UNNLOCRAT   __ __ __ __ __ __ __ __

9. TEW TIAL   __ __ __ __ __ __ __

10. ECSTN SNLAGD   __ __ __ __ __ __ __ __ __ __

11. OINTANIEBRH   __ __ __ __ __ __ __ __ __ __ __

12. LONOYC   __ __ __ __ __ __

13. IANSYR   __ __ __ __ __ __

14. DEYTD EBAR   __ __ __ __ __ __ __ __ __

15. AETTVONSII   __ __ __ __ __ __ __ __ __ __

16. CIEDOTDMEC   __ __ __ __ __ __ __ __ __ __

17. BURHYDNAS   __ __ __ __ __ __ __ __ __

18. ECSXIEER LEHEW   __ __ __ __ __ __ __ __ __ __ __ __ __

19. RAPPEOLS   __ __ __ __ __ __ __ __

CHAPTER 11

Gerbils

1. Fill in the blanks to complete the following sentences.

   (a) Gerbils have a scent gland on the _____, and hamsters have scent glands on their _____.

   (b) Gerbils are active during the _____, and the term for this is _____.

   (c) Hamsters are solitary but gerbils are _____.

   (d) Gerbil pairs are _____, which means they mate for life.

   (e) A _____ is a sudden change in an inherited trait.

   (f) The normal color coat of a gerbil is called _____.

   (g) Another term for *red nose* is _____ _____

   (h) As an alarm to call to others, gerbils _____ with a hind foot.

2. What potential injury could occur if gerbils are provided with an exercise wheel?

   _____

3. Gerbils are member of the family Muridae. What other three other small mammals discussed in your textbook are also members of this family?

   _____ _____ _____

4. Why is it important to never use cedar shavings as bedding material?

   _____

Name _____

5. Complete the Word Scramble relating to the information in Chapter 11.

Word Scramble Gerbils

1. UTNOIMTA   — — — — — — — —
2. OEUGLCAAMF   — — — — — — — — — —
3. RYSETZZ' EAISDES   — — — — — — — -— — — — — — —
4. ALIOGONMN   — — — — — — — — —
5. EBYSTIO   — — — — — — —
6. EARDNTIO   — — — — — — — —
7. UTGOAI   — — — — — —
8. RUSEEZI   — — — — — — —
9. RNGMUIMD   — — — — — — — —

CHAPTER 12

Rats

1. Research on the Internet and write a short paper on the history of the Black Plague. It is also known as bubonic plague, and it devastated hundreds of thousand of people in Europe. Attach this paper to your workbook.

2. Have there been any further outbreaks of the plague? Discuss your findings.

_____

_____

_____

_____

_____

3. A rat that is frightened and defensive displays specific behaviors. What are they?

   (a) _____

   (b) _____

   (c) _____

   (d) _____

   (e) _____

4. Rats are described as being gregarious.

   (a) How does this affect their housing requirements?

   _____

   (b) Which previously discussed small rodent is also gregarious?

   _____

Name _____

5. Complete the Word Scramble for the terms used in Chapter 12.

Word Scramble Rats

1. EDRTNO   — — — — — —
2. URTSAT   — — — — — —
3. TRNUNLACO   — — — — — — — — —
4. IEALG   — — — — —
5. YNPITHRRE   — — — — — — — — —
6. DOIETSCM TAR   — — — — — — —   — — —
7. FASLE   — — — — —
8. P-ANORFGWO   — — — — — — — — —
9. TESANTOGI   — — — — — — — — —
10. ODHEOD   — — — — — —
11. HIEBESKRR   — — — — — — — — —
12. NBILOA   — — — — — —
13. LAGEUP   — — — — — —
14. NPHOLE   — — — — — —
15. CADPE   — — — — —
16. RAIRCSRE   — — — — — — — —
17. EIEGAATRDV   — — — — — — — — — —
18. INNEDRIEGB   — — — — — — — — — —
19. PYLOPAXL SUSINPAOL   — — — — — — — —
                           — — — — — — — —

CHAPTER 13

Mice

1. Four of the most popular small rodent pets have been discussed: hamsters, gerbils, rats, and mice. From the material you have learned, create a chart that compares these four as a potential pet. Your chart should reflect their similarities and their differences in housing, diet, behaviors, and compatibility.

    (a) Which one would you choose for a new pet? _____

    (b) What are the reasons for your choice?

    _____

    _____

    _____

    _____

    _____

2. Access the Web site for the American Fancy Rat and Mouse Association (http://www.afrma.org). Make a list of the categories of information and activities offered by the organization.

    (a) _____

    _____

    _____

    (b) What are the benefits of belonging to a small pet club or organization?

    _____

    _____

    _____

    _____

3. How does the structure of a mouse colony differ from that of other rodent colonies?

    _____

4. What is the most common health problem seen in mice?

   _____

5. Create a set of quiz cards to share with your classmates. Each card should contain one piece of information unique to either a hamster, gerbil, rat, or mouse.

   For example, a quiz card may state, "Some members of these species dance." The answer is "Mouse" (waltzing mice originated in China). Quizzing each other with the cards will help you become more familiar with the differences in these small rodents.

CHAPTER **14**

Guinea Pigs

1. Guinea pigs are very popular as small pets. List the reasons that make them so popular.

   _____

   _____

   _____

   _____

2. Why is it important to limit the amount of alfalfa in the diet of the guinea pig?

   _____

   _____

3. Aside from pelleted foods, which have vitamin C added, which other foods are high in vitamin C and can be fed to a guinea pig?

   _____

   _____

   _____

4. What is the danger of giving guinea pigs certain antibiotics?

   _____

5. Why should a sow have her first litter before the age of 7 months?

   _____

Name _____

6. Complete the Crossword Puzzle relating to the information in Chapter 14.

Crossword Guinea Pigs

Across

3. One reason guinea pigs chatter their teeth
6. Vitamin C
9. Swirls and cowlicks in the coat
10. Variety with short, kinky fur
12. Number of days in the heat cycle of guinea pigs
14. A "bad bite"
15. Cavies do not have this appendage
17. A swelling caused by the accumulation of pus
19. Collective name for a group of cavies
20. A highly developed sense in guinea pigs
21. Some medicines in this group can cause a toxic reaction

Down

1. Contributed to the discovery and production of this serum
2. Term for a male guinea pig
4. Four of the most common external parasites of guinea pigs
5. Bacteria that also causes kennel cough in dogs
7. Means "little pig" in Latin
8. A variety with a rough, wiry coat
11. Type of diet for guinea pigs
13. A condition that may affect sows in late pregnancy
16. Correct term for a guinea pig
18. Term for a female cavy

CHAPTER 15

Chinchillas

1. The Convention on International Trade in Endangered Species (CITES) is an organization that monitors the status of threatened and endangered plants and animals. Before a captive breeding program was established, chinchillas were exploited for their pelts to the point of near extinction. Access the Web site for this organization (http://www.cites.org) and determine the status and population figures for chinchillas in their native habitat. This information may be found in the CITES appendices I, II, and III.

2. Chinchillas have very little ability to defend themselves, but do have one method of escape and one response they use if they feel threatened.

   (a) How do they escape from a predator?

   _____

   (b) What are they able to do when they feel threatened?

   _____

3. What is the importance of providing a dust bath for chinchillas?

   _____

4. How are the different colors coats of chinchillas developed?

   _____

   _____

   _____

5. What is the difference between *enteritis* and *metritis*?

   _____

Name _____

6. Complete the Word Scramble relating to the information in Chapter 15.

Word Scramble Chinchillas

1. DRAIAIG   _ _ _ _ _ _ _
2. OTTIZHEROOP   _ _ _ _ _ _ _ _ _ _ _
3. OTZRGEN   _ _ _ _ _ _ _
4. ITCAHOENPG   _ _ _ _ _ _ _ _ _ _
5. ARPTYMOE   _ _ _ _ _ _ _ _
6. PIIMONCTA   _ _ _ _ _ _ _ _ _

Definitions for Word Scramble:

1. _____
2. _____
3. _____
4. _____
5. _____
6. _____

CHAPTER 16

Ferrets

1. The laws regarding ferrets are sometimes confusing. It may be legal to have a pet ferret in your state, but there may be country or city laws that ban them. Find out if is it legal to keep a pet ferret in the area where you live.

   Yes _____ No _____

   (a) If the answer is no, what are the reasons given?

   _____

   _____

   _____

   (b) If it is legal to have a pet ferret, what are the legal requirements?

   _____

   _____

2. If ferrets are legal in your area, visit the local pet stores and ask where they obtain the kits offered for sale.

   _____

3. Almost all ferrets offered for sale have two blue dots tattooed in one ear. What is the significance of the these tattoos?

   _____

4. Ferrets are delightful and fun pets. If you are familiar with ferrets, describe play behaviors. If not, observe ferrets in a local pet shop and note their behavior with each other. Write your comments in the space below.

   _____

   _____.

5. What is the major medical concern in keeping an unspayed jill? Explain your answer below.

_____

_____

_____

_____

Name _____

6. Complete the Crossword Puzzle relating to the information in Chapter 16.

Crossword Ferrets

Across

2. The most common ferret color

4. Term for a young ferret

6. Severe condition due to the excess production of estrogen

8. Term for a male ferret

9. Stomach problem caused by *Helicobacter*

10. Female hormone

12. Vaccine for ferrets to prevent fatal viral disease

13. The release of eggs from the ovaries

Down

1. Diet fed to ferrets

3. Inflammation of the intestines

5. European relative to the ferret, used in hunting rabbits

7. Genus that also includes weasels, minks, and polecats

11. Term for a female ferret

CHAPTER 17
Amphibians

1. Amphibians have a life cycle that begins in the water; they then undergo physical changes called metamorphosis that enable them to live on land. In the space below, draw the typical life cycle of an amphibian.

2. Give two reasons why a handler should wear latex gloves when handling an amphibian.

   (a) _____

   (b) _____

3. Describe three different habitats required to meet the needs of these amphibians:

   (a) African Clawed Frog _____

   (b) Tiger Salamander _____

   (c) Common Leopard Frog _____

4. What are two reasons it is important to provide clean, dechlorinated water for all amphibians?

   (a) _____

   (b) _____

5. Which species is most affected by red leg? Discuss what it is, what causes this condition, and how it is prevented.

   _____

   _____

   _____

Name _____

6. Complete the Word Scramble relating to the information in Chapter 17.

Word Scramble Amphibians

1. UTACADE   — — — — — — —
2. RANOUEOSM   — — — — — — — — —
3. SPEOAPRORHTME   — — — — — — — — — — — — — —
4. SOSSOIM   — — — — — — —
5. ARALV   — — — — —
6. SEAATILIN   — — — — — — — —

Definitions for Word Scramble:

1. _____

2. _____

3. _____

4. _____

5. _____

6. _____

CHAPTER 18
Reptiles

1. How is a reptile heart different from a mammalian heart, and what effect does this have?

   _____

   _____

2. What are two of the basic differences between turtles and tortoises?

   _____

   _____

3. Give three reasons why UVB light needs to be provided for captive reptiles.

   (a) _____

   (b) _____

   (c) _____

4. If a snake is having difficulty shedding, what things can be done to assist the snake?

   _____

   _____

5. What is the importance of knowing the POTZ range for specific reptiles?

   _____

   _____

Name _____

6. Complete the Crossword Puzzle relating to the information in Chapter 18.

Crossword Reptiles

Across

2. Correct term for "cold-blooded"
6. What reptiles do to raise their body temperature
7. The lower part of a chelonian shell
10. Taxonomic family with the largest species of snakes
11. Bacterial disease that can be transmitted to humans
12. Living in trees
13. Eggs that hatch internally
14. The upper part of a turtle shell

Down

1. Visual differences between the senses
3. The transparent eye covering
4. The adhesive pads on the feet of geckos
5. Number of chambers in a reptile heart
8. Eggs that hatch externally
9. A habitat that replicates a natural environment
12. The smallest member of the iguana family

CHAPTER 19

Birds

1. List the anatomical features that give birds the ability to fly.

   (a) _____

   (b) _____

   (c) _____

   (d) _____

   (e) _____

   (f) _____

2. In the space provided below, draw the beak and feet of a Psittcine, and by comparison, the beak and feet of a Passerine.

3. Without referring to your textbook, label the female reproductive system.

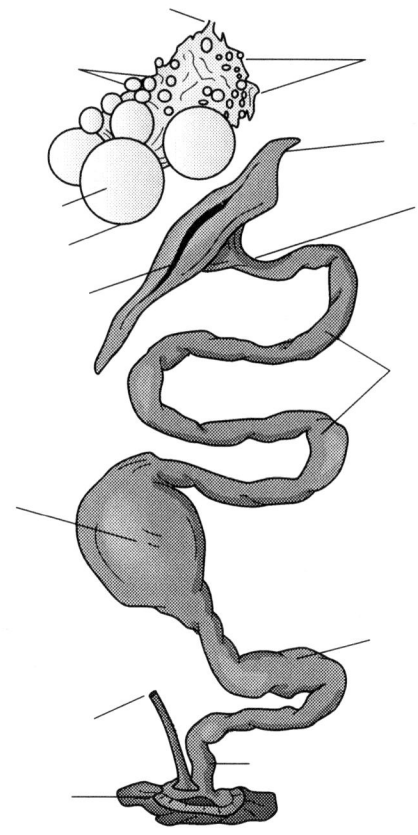

4. What is unique about the tongue of a lorry?

_____

5. Which bacterial disease of birds can be transmitted to humans? Discuss the disease, the causative agent, and how it can be transmitted to people.

_____

_____

_____

_____

_____

Name _____

6. Complete the Crossword Puzzle relating to the information in Chapter 19.

Crossword Birds

Across

1. Sexually dimorphic bird with male and female having different colors
4. A place where birds are housed
6. The fleshy part of a bird's beak
7. The feathers that determine the shape of the bird
8. Junction of the urinary, digestive, and reproductive tracts
9. Popular pet bird often called a parakeet
13. The perching birds
15. Describing the bones of birds
17. An enlargement of the esophagus that stores food
19. The keel bone
20. A food that needs to be provided for lories
21. The last section of the spine, the tail area
22. The largest genus group

Down

2. A subspecies of the African Gray
3. The largest organ in the digestive system of a bird
4. The first bird
5. A painful ailment of the feet
10. A condition caused by an iodine deficiency
11. Anatomical part where the shell is formed
12. When birds groom themselves
14. The thick, white substance secreted around the yolk
16. The largest of all the macaws
18. A cockatiel that is primarily yellow

CHAPTER 20

Fish

1. Many aquarists have kept piranha fish; however, many owners have been irresponsible. As a consequence, it is now illegal in many places to buy or possess these fish.

   (a) What actions prompted the banning of these fish?

   _____

   (b) Are piranhas legal in your area? Yes _____ No _____

   (c) What species is sometimes sold as a "false piranha"?

   _____

   You will need to research your answer by visiting a local aquarium or pet store.

2. What is the rule of thumb that determines the number of fish that can be put into a freshwater tropical aquarium?

   _____

   What is different about this rule if it is a saltwater aquarium?

   _____

3. Calculate the inches of fish that could be housed in a freshwater tank that measures 24 inches × 18 inches × 12 inches.

   Show your method of calculation and answer: _____

   _____

4. What instrument measures the degree of salinity?

   (a) _____

   What are two other terms for this same measurement?

   (b) _____

5. Label the internal anatomy of a fish.

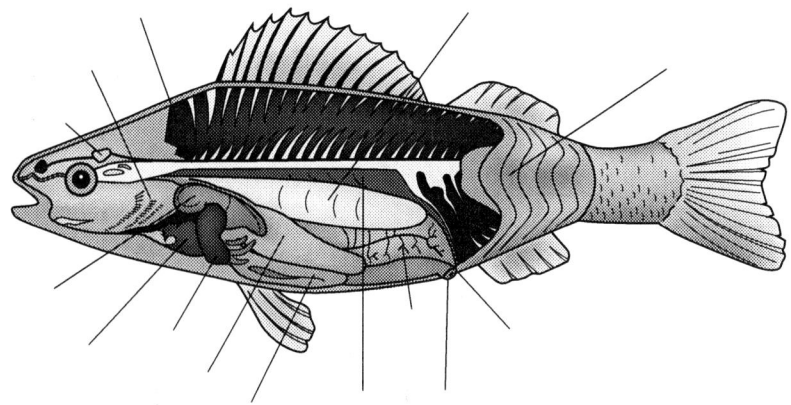

Name _____

6. Complete the Crossword Puzzle relating to the information in Chapter 20.

Crossword Fish

Across

4. Pigment cells that give fish their color

8. Contributes to an increase of ammonia and nitrates

10. Small crustaceans called "water fleas"

12. Filtration system that uses bacteria

13. Fish sperm

14. Reproduction ritual for egg-layers

15. Whisker-like projections around the mouth

Down

1. pH of 0

2. The modified anal fin of some male fish

3. An ornamental species of carp from Japan

5. A mutually beneficial relationship

6. Term for newly hatched or newborn fish

7. A critical factor in most tanks

9. The pressure-sensitive cells within the lateral line

11. The exchange of oxygen and carbon dioxide in the water

CHAPTER 21

Hedgehogs

1. Why do certain species tend to become "fad pets"? Consider the contributing factors and write your answer in the space provided.

   _____

   _____

   _____

   _____

2. A suggested activity in your textbook is to research some of the diseases of hedgehogs. Find what information you can regarding a condition called *wobbly hedgehog syndrome.*

   What is it? _____

   _____

   What are the early signs of this condition?

   _____

   _____

   What is the outcome of this condition?

   _____

3. Discuss with a veterinarian the best way to treat a mite infestation in a pet hedgehog and write his or her response in the space below.

   _____

   _____

4. If the hedgehog is not exposed to other hedgehogs, how do mites become a problem?

   _____

Name _____

5. Complete the Word Scramble relating to the information in Chapter 21.

Word Scramble Hedgehogs

1. DARAIRHE   — — — — — — — —
2. ITRIAPRTONU   — — — — — — — — — — —
3. SNSPIE   — — — — — —
4. NERVIOCESTI   — — — — — — — — — — —
5. GONITNANI   — — — — — — — — —
6. IKNTRAE   — — — — — — —
7. THLOGE   — — — — — —
8. .A NEVRUASILBT   — — — — — — — — — — — — —

Definitions for Word Scramble:

1. _____
2. _____
3. _____
4. _____
5. _____
6. _____
7. _____
8. _____

CHAPTER 22

Sugar Gliders

1. Without even handling a sugar glider, what is the easiest way to determine the sex?

   _____

2. Refer to Figure 22–2. Is this a male or female sugar glider? Male _____ Female _____

3. Many popular exotic species are not legal to keep as pets in some areas. Contact your local animal control office to see if sugar gliders can be kept as pets in your area. If they are not legal, ask the reason(s) they are banned.

   Write the response to your questions below.

   _____

   _____

   _____

   _____

4. What potential zoonotic diseases could be transmitted to people from sugar gliders?

   _____

   _____

5. Research the Internet for other species with syndactylism. What do they have in common?

Name _____

6. Complete the Word Scramble relating to the information in Chapter 22.

Word Scramble Sugar Gliders

1. OCAACL   — — — — — —

2. MUG   — — —

3. SECNT NDALG   — — — —   — — — — —

4. NYASDLTCY   — — — — — — — —

5. ONYCSIMCAITSO   — — — — — — — — — — — —

6. URNOANCLT   — — — — — — — —

7. SPA   — — —

8. SARUPIMUM   — — — — — — — —

9. IAACAC   — — — — — —

10. RIPLSMAAU   — — — — — — — —

11. MGPUAITA   — — — — — — — —

12. UHCOP   — — — — —

13. EMNIOVORS   — — — — — — — — —

14. IRSAMIPP   — — — — — — — —

15. JESOY   — — — — —